Collins

INTERNATIONAL
LOWER SECONDARY

Computing
Workbook

7

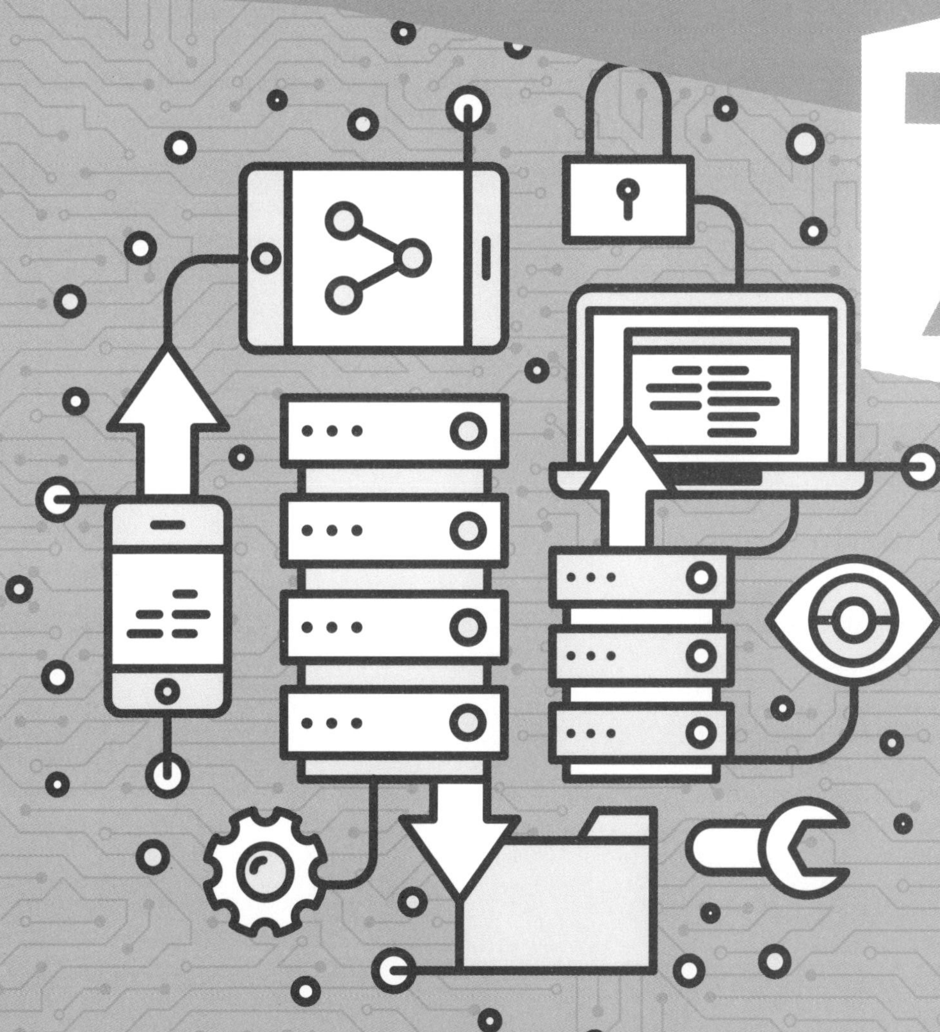

Rebecca Franks, Dr Tracy Gardner and Liz Smart

William Collins' dream of knowledge for all began with the publication of his first book in 1819.

A self-educated mill worker, he not only enriched millions of lives, but also founded a flourishing publishing house. Today, staying true to this spirit, Collins books are packed with inspiration, innovation and practical expertise.

They place you at the centre of a world of possibility and give you exactly what you need to explore it.

Collins. Freedom to teach.

Published by Collins

An imprint of HarperCollins*Publishers*
The News Building, 1 London Bridge Street, London, SE1 9GF, UK

HarperCollins*Publishers*
Macken House, 39/40 Mayor Street Upper, Dublin 1, D01 C9W8, Ireland

Browse the complete Collins catalogue at
collins.co.uk

10 9 8 7 6 5 4 3 2 1

ISBN 978-0-00-868405-1

British Library Cataloguing-in-Publication Data

A catalogue record for this publication is available from the British Library.

Authors: Rebecca Franks, Dr Tracy Gardner and Liz Smart
Publisher: Catherine Martin
Product manager: Saaleh Patel
Project manager: Just Content Ltd
Development editor: Julie Bond
Copy editor: Becca Law
Proofreader: Laura Connell and Tim Jackson
Illustration: Jouve India Ltd
Cover designer: Gordon McGilp
Cover image: Amparo Barrera, Kneath Associates
Internal designer: Steve Evans, Planet Life Art
Typesetter: Ken Vail Graphic Design
Production controller: Lyndsey Rogers
Printed and bound by Martins the Printers

MIX
Paper | Supporting responsible forestry
FSC www.fsc.org FSC™ C007454

This book contains FSC™ certified paper and other controlled sources to ensure responsible forest management.

For more information visit: harpercollins.co.uk/green

Acknowledgements

Support materials and screenshots are licensed under the Creative Commons Attribution-ShareAlike 2.0 license. We are grateful to the following for permission to reproduce screenshots. In some instances, we have been unable to trace the owners of copyright material, and we would appreciate any information that would enable us to do so.

Python Software Foundation: Permission obtained to use screenshots demonstrating Python programming language features.

Scratch Foundation: Authorised usage of screenshots showcasing Scratch programming environment elements.

EduBlocks: Granted clearance for the inclusion of screenshots depicting EduBlocks interface and functionalities.

Scratch is developed by the Lifelong Kindergarten Group at the MIT Media Lab: p.6, p.9.

PSF's License Agreement and PSF's notice of copyright, i.e., "Copyright (c) 2001 Python Software Foundation; All Rights Reserved" are retained in Python 2.0.1 alone or in any derivative version prepared by Collins (Licensee): p.50, p.52, p.54–55, p.57–58.

EduBlocks: p.50, p.53–54.

Images

p.7 Pavel Vinnik/ Shutterstock

Contents

Introduction

The Collins Stage 7 Student's Book and Workbook offer a rich programme of skills development, based on a varied and stimulating set of projects grounded in real-world contexts.

The Stage 7 Workbook directly supports the Student's Book, with Workbook tasks for every lesson in the Student's Book. The Workbook tasks are clearly referenced in the Student's Book to make the link between the two resources clear.

Each task provides the context and scaffolding required to complete any paper-based tasks during a lesson. This allows you to build a record of your notes and designs that are separate to the work you create using software. You can use the Workbook to remind yourself about plans and designs you have created and concepts you have learned about.

Regular reflection tasks are included in which you have an opportunity to privately consider your thoughts about a particular lesson or a skill you have been practising. These can be shared with the class if there is an opportunity or kept private between you and your teacher.

Each lesson ends with an optional homework task in the Workbook which your teacher may set to help you consolidate what you have learned during the lesson, or research something new to enhance your learning about the topic.

We hope our approach helps you and your teachers to enjoy learning about computing, digital literacy and ICT in a practical and enjoyable way.

Rebecca Franks, Dr Tracy Gardner and Liz Smart

Chapter 1.1 Safe and healthy use of technology

Task A. Protecting your wellbeing online

What actions do you take to protect your wellbeing online? If you need ideas, refer to the tips in the Student's Book.

Key terms

Wellbeing – The way people feel about themselves and their lives

....

....

....

....

....

Return to page 3 of the Student's Book.

Task B. Stop and think!

Have you seen pop-ups like these before? What would you click on? Why?

Key terms

Permission – When a website or app asks to access your data or sensors like camera or microphone

Cookie – Small file of information that a website stores on your computer

Policy – Document explaining what you can expect from a website or service, including how it will use your data

	Popup	Your experience
1	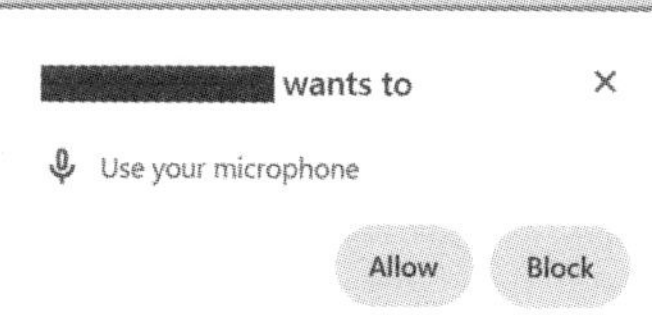	
2	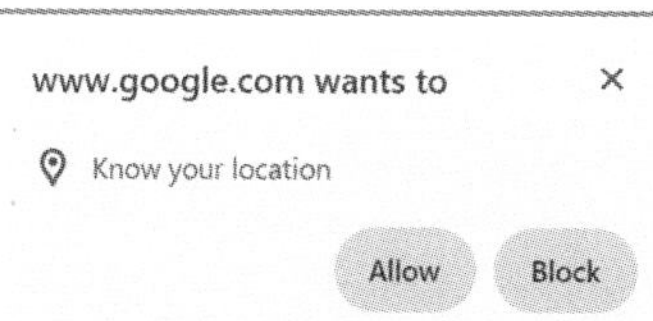	
3		
4		

Return to page 5 of the Student's Book.

Task C. Practice scenarios

Discuss each scenario below. Make notes on how each scenario involves wellbeing, taking action and/or personal responsibility.

	Scenario	Discussion
1	Farah has made a friend online who she is planning to meet up with. She has decided that when she does meet them she will make sure she is not alone.	
2	Ali sees an unkind comment on a project on the Scratch website. There is a report button available.	
3	Manon always clicks on 'Accept all cookies' because it's quick.	
4	Jay sees a chat message where Ali has made an unkind comment about Eva.	
5	Afzal has found an exciting website that shows teenagers climbing up scaffolding on building sites.	
6	You get a pop-up on a website requesting access to your camera.	

Return to page 5 of the Student's Book.

Reflection

How will you act differently online after this lesson?

....

....

....

Homework

- Discuss today's lesson with a trusted adult.
  - Do they have any techniques they use for looking after their wellbeing when using online tools?
  - Have they ever had to take action because of cyberbullying?

Chapter 1.2 Responsible participation in online communities

Task A. Dealing with trolling behaviour

It is not your fault if you encounter trolling behaviour, but there are actions that you should and shouldn't take to improve the situation for yourself and others.

Key terms

Trolling – Deliberately unhelpful behaviour in an online community

Underline the correct word. When you encounter a troll:

1 You [should / shouldn't] try to convince them to be nice.

2 You [should / shouldn't] ignore them.

3 You [should / shouldn't] be mean to them.

4 You [should / shouldn't] report them.

5 You [should / shouldn't] ask an adult for support if the behaviour upsets you.

Return to page 6 of the Student's Book.

Task B. Online community guidelines

You are going to write a set of guidelines for an online community of your choice.

1 Agree a community topic with your group and write it here:

Community topic: ....

2 Agree 4–8 guidelines to make sure that community members have a positive experience.

....

....

....

....

....

....

....

....

Return to page 7 of the Student's Book.

Reflection

What guidelines did you agree? How difficult would they be to follow?

....

....

....

....

....

....

Homework

Find and read through the community guidelines for the Raspberry Pi Foundation and Tynker.

> **Tip** Your teacher will provide you with links to the relevant web pages.

Summarise your findings.

....

....

....

....

....

....

....

....

Chapter 1.3 Applications of AI

Task A. AI definitions

Write definitions of these terms in your own words, giving examples.

1 Artificial intelligence

....

....

....

....

2 Artificial general intelligence

....

....

....

....

Return to page 9 of the Student's Book.

Task B. AI in your life

Complete the table to show examples of AI that you interact with in your everyday life.

Example	**Type of AI** (voice assistant, predictive text, image recognition, recommendation system, or others)

Key terms

Voice assistant – AI system that can perform useful actions in response to voice commands

Recommendation system – Computer system that makes predictions on content that a user would like, using AI

Image recognition – Identification of objects in images by computer systems using AI

Return to page 10 of the Student's Book.

Task C. How do you learn to play a computer game?

Type in or click on the link that your teacher provides.

Click the green flag and press the space bar to launch the launcher with the right amount of power to stop the message reaching the troll. You may prefer to play in Full Screen mode.

Make notes as you learn how to play 'Do not feed the troll'. You can use words and/ or diagrams.

How do you decide when to press the space bar?

What data are you using?

Return to page 10 of the Student's Book.

Reflection

How did you learn to play the game?

....

....

....

....

Homework

Play the role of a machine learning plant identification app. Look at the plants you pass on the way home. What similarities and differences do they have?

Similarities: ....

Differences: ....

Do you get faster at identifying plants when you see them often? Why do you think this is?

....

....

Chapter 1.4 Train a machine learning model

Task A. Complete the AI definitions

Complete the following sentences using the keywords you have learnt:

Key terms
Machine learning – Type of AI that uses data and algorithms to gradually produce more accurate results in a way that imitates how humans learn

1 A machine learning .... combines an algorithm and .... data.

2 .... machine learning is when a human adds .... to data.

3 .... machine learning is when a computer detects .... in data.

Task B. Labelling data

Name some objects that the machine learning model for a self-driving car would need to be able to recognise:

1 ....

2 ....

3 ....

4 ....

5 ....

Return to page 11 of the Student's Book.

Task C. First training data

Your first training data should look similar to this:

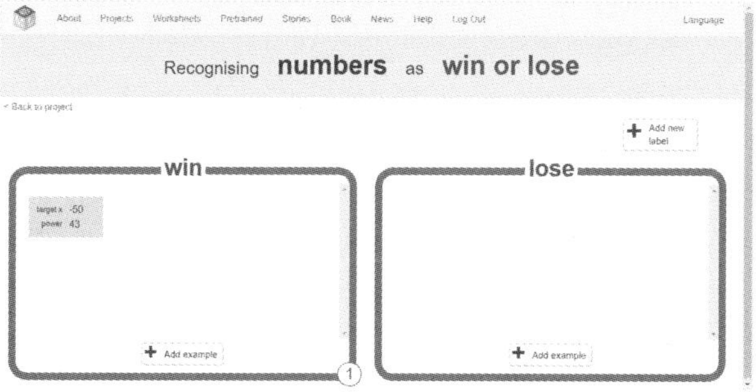

Record your first training data. Was it a win or a lose?

win lose

Return to page 13 of the Student's Book.

Task D. After 20 attempts

How well is your model playing the game after at least 20 attempts? (It is a mixture of your playing and random playing.)

The total number of wins and the total number of loses from the training data are shown in the circles at the bottom right of each box.

Wins: .... Loses: ....

Describe in words how well the game is playing. Is it better than random? Is it better than you?

....

Return to page 14 of the Student's Book.

Task E. After many attempts

When your teacher asks you to record your final values, fill in the total number of wins and the total number of loses in the training data.

Wins: .... Loses: ....

Describe in words how well the game is playing. Is it better than random? Is it better than you?

....

Return to page 15 of the Student's Book.

Reflection

How did the training data improve the AI?

....

....

....

Homework

Your teacher will give you a project from zooniverse.org to explore.

Tip You don't need to sign up, but you should practise identifying images.

Name of project: ....

Summarise your findings.

....

....

....

Chapter 1.5 Make your own machine learning project

Task A. Plan your project

Open Scratch to get inspiration. You can modify costumes or create your own if you have time.

Remember: The player controls the power of the 'Launcher'.

Sprite/Backdrop	Role	Your choice
	Target	
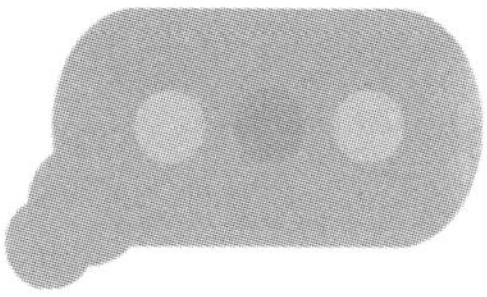	Projectile	
	Launcher	
	Backdrop/Colour	

Return to page 16 of the Student's Book.

Task B. Record your results

Play until the machine learning model loses. Observe the '*x* position' and 'power'. Why do you think the machine learning model made a mistake?

x position: ....

power: ....

Reason: ....

....

....

Approximately how many turns (wins + loses) does it take before the AI is playing better than you?

....

How many rounds can the AI win without losing? ....

How many wins and loses does your training data have at the end of the lesson?

Win: ....

Lose: ....

Return to page 19 of the Student's Book.

Reflection

What would happen if you changed the game by moving the Target to a different position or changing the starting position of the Projectile?

....

....

....

Homework

Discuss with a trusted adult: 'Could a computer be taught to play any game?'

What conclusion did you come to?

....

....

....

....

Chapter 1.6 Share findings on AI and machine learning

Task A. Record your findings

Summarise your results from the last two lessons.

....

....

....

....

Summarise the key findings of your group.

....

....

....

....

Return to page 20 of the Student's Book.

Task B. Reflection

Summarise what you have learnt about artificial intelligence and machine learning.

What is artificial intelligence?

....

....

What is artificial general intelligence?

....

....

What is machine learning?

....

What is machine learning good for?

....

....

What is training data?

....

....

Why is it important to have the right training data?

....

....

Return to page 20 of the Student's Book.

Reflection

What kinds of problems is machine learning good for?

....

....

....

Project skills checklist

- ☐ I can plan a machine learning project
- ☐ I can choose costumes and a backdrop in Scratch
- ☐ I can modify a Scratch project based on my design decisions
- ☐ I can train a machine learning model
- ☐ I can investigate the effectiveness of my trained model
- ☐ I can share my findings

Homework

Identify three problems that would be best solved by machine learning.

1 ....

....

2 ....

....

3 ....

....

Chapter 2.1 Collaborative working

Task A. Breaking news!

A source has sent you a message about a penguin that has escaped from a conservation centre. They have managed to interview a couple of people who saw the escape.

Escaped Penguin

Hi, I've just found out that a penguin got out of their enclosure and wandered out of the conservation centre. I managed to speak to a couple of witnesses. This is what they said:

"I saw the penguin waddling down the road but thought someone was just playing with a remote controlled penguin toy – I just didn't imagine a penguin could escape!" – A visitor to the centre

It's chaos here, this has never happened before. A search party is gathering, they are grabbing blankets, buckets of fish and ice. Nobody knows how to find a penguin" – A centre veterinarian

There is a hole at the back of the enclosure and that is how the escape began. I believe the penguin has been seen in the centre of town but that is unconfirmed. Can you write a newspaper article about it please? People need to know.

Create a collaborative news article for the local newspaper. You will need to grab the attention of your audience with a headline, key facts, quotes and formatting.

Tip Typing tip: Remember to use the raised homing bars or dots on your keyboard to position your hands in the most efficient way when typing.

Space for notes.

....

....

....

....

....

....

....

....

Return to page 24 of the Student's Book.

Task B. Co-author reflection

What do you think worked well when collaborating?

....

....

....

What would you do differently next time?

....

....

....

My top three ground rules for collaborative working in a word processor are:

1 ....

2 ....

3 ....

Return to page 25 of the Student's Book.

Reflection

If all of your collaborative work could be completed online from anywhere with a computer, why would/wouldn't you want to do it that way all of the time?

....

....

....

Homework

Watch or listen to a local news report on the television or radio.

1 What would the headline be for the news report?

....

2 What are three key facts from the news report?

....

....

....

Chapter 2.2 Usage rights

Task A. Image usage rights

1 Use a search engine to find out about different image licence types and then summarise your findings.

a Royalty free

....

....

....

b Creative Commons and attribution (Creative Commons BY)

....

....

....

c Copyright free

....

....

....

2 List other advanced search tools you could use to filter your images:

....

....

....

....

3 Practise searching for these images and restricting the results by licence type and other filters. Tick the box when you have found a suitable image.

a Mountains ☐

b Seahorse ☐

c Storm clouds ☐

d Concert ☐

4 For each of these descriptions, search for and select an image that you are able to use and then insert it into a document. Tick the box when you have found and inserted a suitable image.

a World record breaker ☐

b Best place to visit ☐

c Yellow photography ☐

d 3D art ☐

Return to page 29 of the Student's Book.

Reflection

Did you find images that you wanted to use but weren't allowed to? How did you feel about this?

....

....

....

Homework

Interview someone from an older generation to you. It could be an adult in your household or a member of your local community.

Ask them how they heard about the latest news when they were younger.

Write down their response below:

....

....

....

Chapter 2.3 Future technologist reliability

Task A. Researching predictions

1 Write down keywords or phrases you could use to find out some recent:

a current technology trends

....

....

....

....

b future technology predictions.

....

....

....

....

Use your keywords to search the internet. Write about the predictions you find.

Tip Include source details and the date the prediction was made.

....

....

....

....

....

....

....

....

....

....

Return to page 30 of the Student's Book.

Task B. Establish trust

Use the websites you found earlier to investigate how trustworthy the predictions are.

Remember to think about:

- How often have they created content?
- Is the content or author sponsored or biased?
- What are the author's credentials?
- How many views has the content had?
- Are there useful reviews about the author?
- Have they used citations for sources?
- Do they pass a plagiarism checker?

1 I think these predictions are trustworthy because:

....

....

....

....

....

....

2 I think these predictions are untrustworthy because:

....

....

....

....

....

....

3 Based on my research and my knowledge of current technology trends, I think the prediction that is most likely to come true is:

....

....

Return to page 31 of the Student's Book.

Reflection

Write down the prediction that is most likely to happen, based on the research you have carried out.

....

....

....

Homework

Choose a piece of technology that is familiar in your household, for example, a microwave oven or a toaster. Make a prediction about how this technology could change in the future.

Which technology did you choose?

....

What is your prediction?

....

....

....

Chapter 2.4 Plan a news report

Task A. News report overview

Work with your group to agree on roles for everyone.

The roles for our news report are:

Role	Person

Consider your trustworthy predictions from the last lesson as a starting point to help you to agree on which future technologies to discuss.

The future technologies prediction we will discuss is:

....

....

....

Return to page 32 of the Student's Book.

Task B. Interview observations

As you rewatch the video, answer the following questions:

1 What is the overall structure of the news report?

....

....

2 How does the presenter introduce the interview and the experts?

....

....

3 Do the people on camera have the script written down, memorised or were they using off-camera prompts?

....

....

4 Which camera angles are used?

....

....

5 Does the news report use any props, diagrams or graphics?

....

....

6 How does the news report end?

....

....

Return to page 33 of the Student's Book.

Task C. News report notes

Use your ground rules for collaborative working to plan how to write your script. Make notes on the next page about your responsibilities when writing the script.

Write down any props you will need to bring or make for the recording in the next lesson.

Tip Typing tip: Sitting correctly at the computer will help you improve your typing speed and accuracy, and make sure you avoid strain to your wrists and hands. Look directly at the computer and keep your back straight. Your feet should be flat on the ground and your wrists in a natural position.

....

....

....

....

....

....

....

....

....

....

Return to page 33 of the Student's Book.

Reflection

What were the advantages and disadvantages of working collaboratively on the script?

....

....

....

Homework

Find a local news article and practise reading it out loud to a pretend audience or a member of your household. This will help you develop your presentation skills for next lesson.

1 Which news article did you choose?

....

2 Did practising help to improve your presentation skills? Explain your answer.

....

....

....

Chapter 2.5 Record a news report

Task A. Using the video recorder

Write notes to help you to use the video recording equipment and transfer the final file.

....

....

....

....

....

....

....

....

....

....

....

....

....

....

....

....

....

....

....

....

Return to page 35 of the Student's Book.

Task B. Video notes

Write notes of things to remember, or improve, during future takes.

Task C. My video

1 The file size of my final video is ....

2 The place my video is saved is ....

Return to page 35 of the Student's Book.

Reflection

Why would it have been easier to record the video in more than one take and then edit it together later?

....

....

....

Homework

The video that you recorded today may have taken several takes. It also might need to be cut into smaller segments to remove unwanted pauses or words. Watch a local TV news report and mark when you spot these key editing features:

- ☐ The camera angle changes to focus on different people.
- ☐ There is additional video of scenes related to the news report.
- ☐ The reporter adds narration to the additional video scenes (for example, you cannot see their face when they are talking).
- ☐ Background music or sound effects have been added.
- ☐ There is text to show the name of the news reporter or any interviewees.

What was the topic of the news report that you watched?

....

Chapter 2.6 Broadcast a news report

Task A. Reflection

1 How likely is it that your future technology prediction will become a reality?

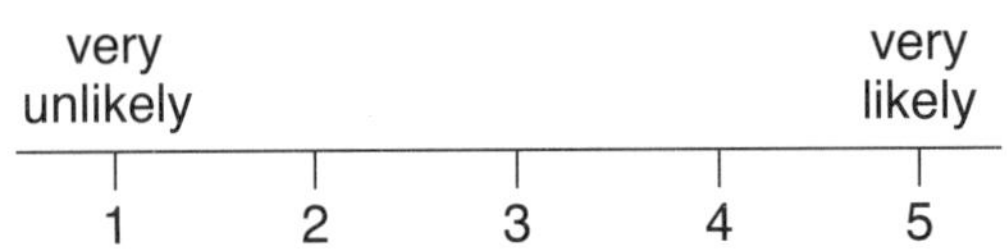

2 Why did you give this score?

3 What was your favourite part of your online collaboration?

4 Which part of your online collaboration would you improve if you had more time?

5 When you were viewing the news reports that other groups created, did anything stand out as an improvement you'd like to make to your news report?

Return to page 36 of the Student's Book.

Reflection

Share something you really liked about a news report video made by someone else.

....

....

....

Project skills checklist

- ☐ I can design a news report
- ☐ I can make predictions about future technologies
- ☐ I can write a script
- ☐ I can develop a presentation style
- ☐ I can record a news report
- ☐ I can work collaboratively with others

Homework

Write down three things that you predict will happen with technology in the future.

1 ....

....

2 ....

....

3 ....

....

Chapter 3.1 Logic gates

Task A. Boolean expressions

Circle the images that would result in a TRUE outcome from the Boolean expressions.

1 'Black AND square'

2 'Circle OR black'

3 'NOT circle'

4 'Black OR NOT square'

Key terms

Boolean logic – Form of algebra that uses the operators AND, OR and NOT

Boolean expression – Expression that either has a TRUE or FALSE outcome

Logic gates – Used in logic circuits in computers. They are based upon the Boolean logic operators AND, OR and NOT

Return to page 40 of the Student's Book.

Task B. Investigate logic gates

Your teacher will show you how to open the 'Logic Gates' Scratch program. Use it to answer the questions in this task.

Tip The output of each gate will be a purple dashed line for ON or a grey line for OFF.

Click the 'Full Screen' option in Scratch and then press the green flag to start the program.

For each question, circle the correct answer: either ON or OFF.

1 Operate the AND gate. Switch ON the first input and switch ON the second input. What is the output?

ON or OFF

2 Still on the AND gate. Switch OFF the first input and keep the second input ON. What is the output?

ON or OFF

3 Still on the AND gate. Switch ON the first input and switch OFF the second input. What is the output?

ON or OFF

4 Move to the OR gate. Switch ON the first input and switch ON the second input. What is the output?

ON or OFF

5 Still on the OR gate. Switch OFF the first input and keep the second input ON. What is the output?

ON or OFF

6 Still on the OR gate. Switch ON the first input and switch OFF the second input. What is the output?

ON or OFF

7 Move to the NOT gate. Switch ON the input. What is the output?

ON or OFF

8 Still on the NOT gate. Switch OFF the input. What is the output?

ON or OFF

Return to page 41 of the Student's Book.

Reflection

Write down the Boolean expression that you shared with a partner.

....

....

Homework

Find and write down examples of Boolean expressions from everyday life. They could be rules you have in your home, instructions for playing a game or statements about the weather.

....

....

Chapter 3.2 Selection in flowcharts

Task A. Label the flowchart symbols

Draw lines to match the labels with their flowchart symbols.

1.

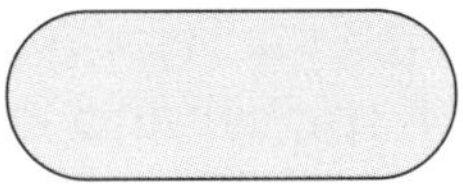

2.

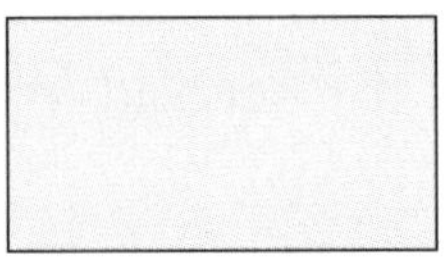

3.

4.

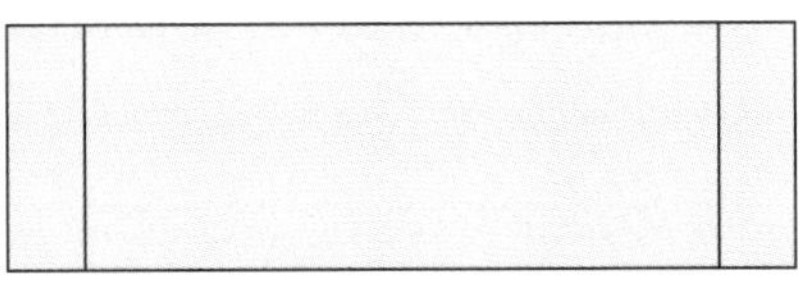

5.

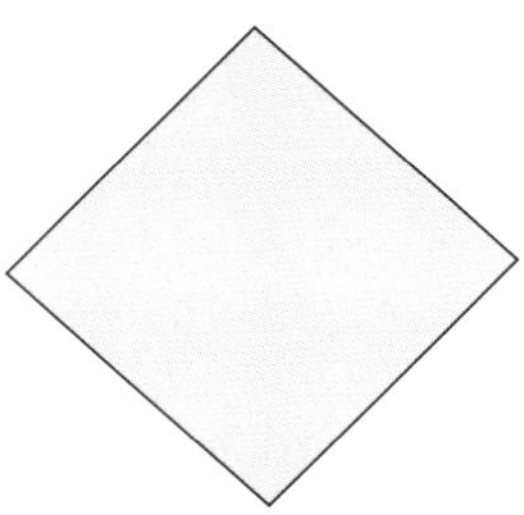

A. Process

B. Connector

C. Decision

D. Start / Stop

E. Subroutine (procedure)

Return to page 42 of the Student's Book.

Task B. The correct ingredients

Complete the table below to predict the output of the flowchart based on different inputs. The first one has been completed for you.

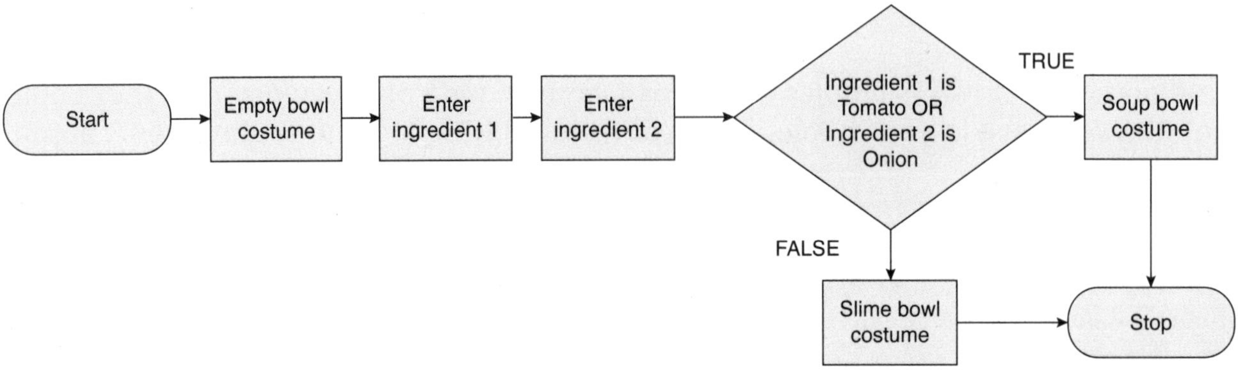

Ingredient 1	Ingredient 2	Output
Tomato	Onion	Soup bowl costume
Tomato	Potato	
Potato	Onion	
Carrot	Potato	
Carrot	Tomato	

Return to page 43 of the Student's Book.

Reflection

Were any of your predictions incorrect? If so, which ones and what was the actual output?

....

Homework

Create a new recipe condition for your own soup recipe. Remix and modify the Scratch project to use your new condition.

Fill in the decision symbol with the condition for your recipe.

Chapter 3.3 Debug flowcharts

Task A. Debug the flowcharts

Three flowcharts have been created for different purposes. For each flowchart:

1 Read the description for how the flowchart is expected to run.

2 Circle the place where the error occurs.

3 Redraw the flowchart with the error fixed.

Flowchart 1 – The lighthouse

A student has designed a lighthouse project in Scratch. It should ask the user if it is night-time or stormy. The lighthouse should display a rotating light if the condition is TRUE. Circle the error.

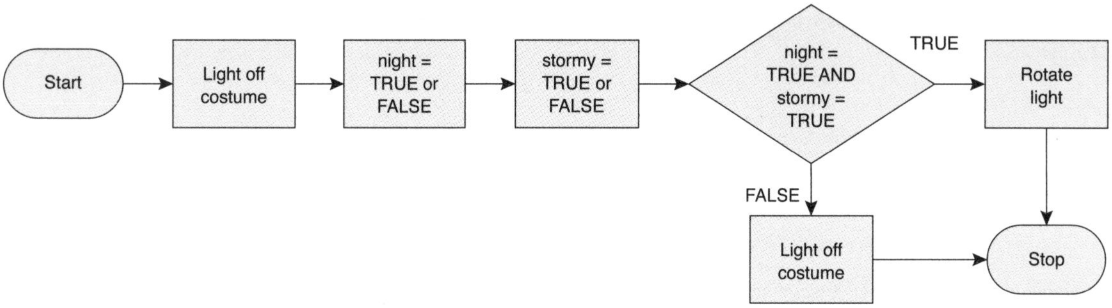

Redraw the flowchart with the error fixed.

Flowchart 2 – Intruder alarm

When the alarm is switched on and motion is detected, the alarm sound should play.

Circle the errors.

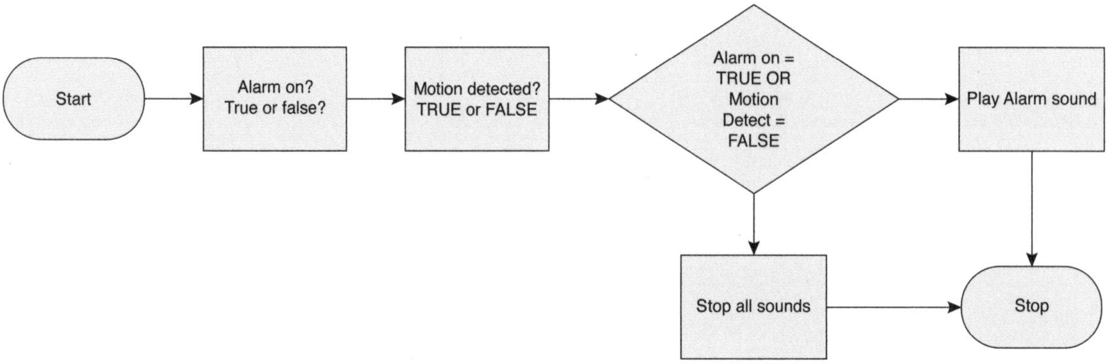

Redraw the flowchart with the error fixed.

Flowchart 3 – Climate control

If the temperature is over 25 degrees Celsius then the air conditioner should switch on and the heater should be off. If the temperature is below 10 degrees Celsius then the heating should switch on and the air conditioning should be off. If the temperature is between 10 and 25 then the heater and air conditioner should be switched off.

Circle the errors.

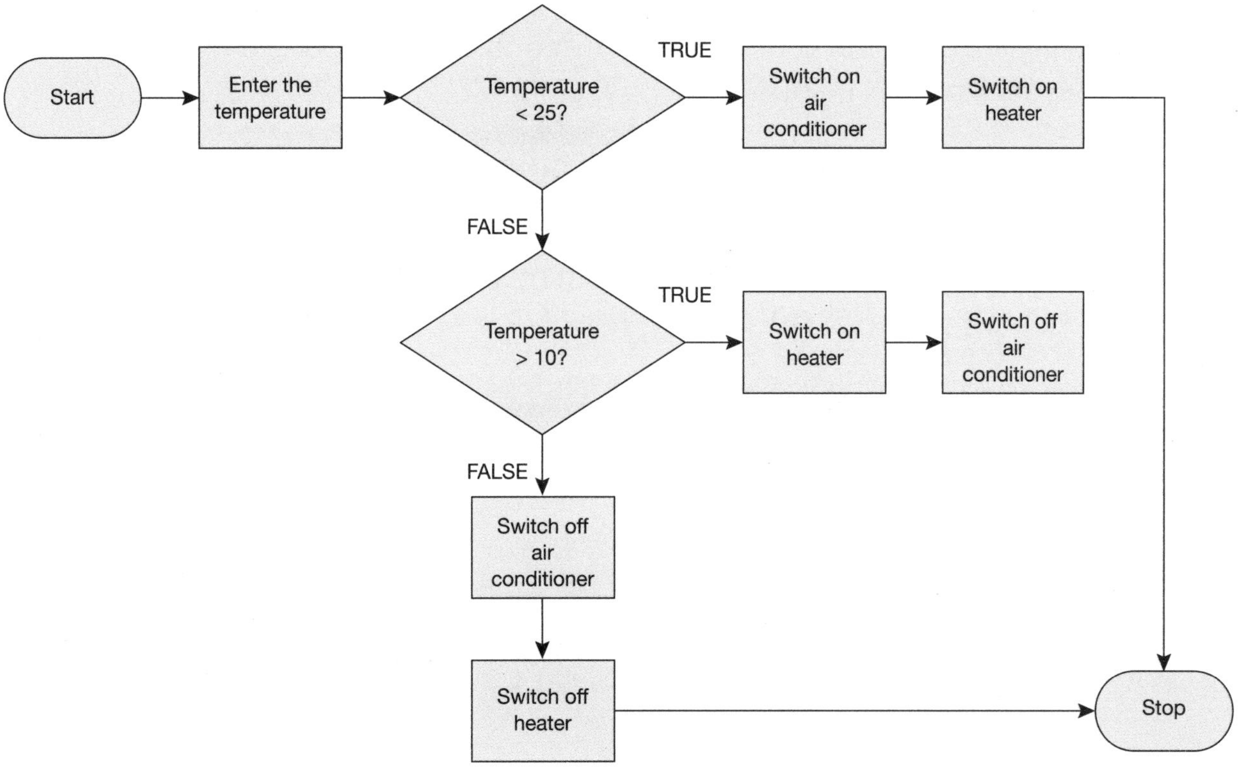

Redraw the flowchart with the errors fixed.

Return to page 44 of the Student's Book.

Reflection

The flowcharts in the Student's Book will only check for a condition once and then the program will stop. What could be introduced to the flowchart to make the program constantly check for changes in the condition?

....

Homework

A student has created a flowchart for a Scratch project for creating a salad. The user needs to enter 'Lettuce' as the first ingredient and 'Cucumber' as the second ingredient to successfully create a salad.

There are two mistakes in the flowchart. What are the mistakes and how would you fix them?

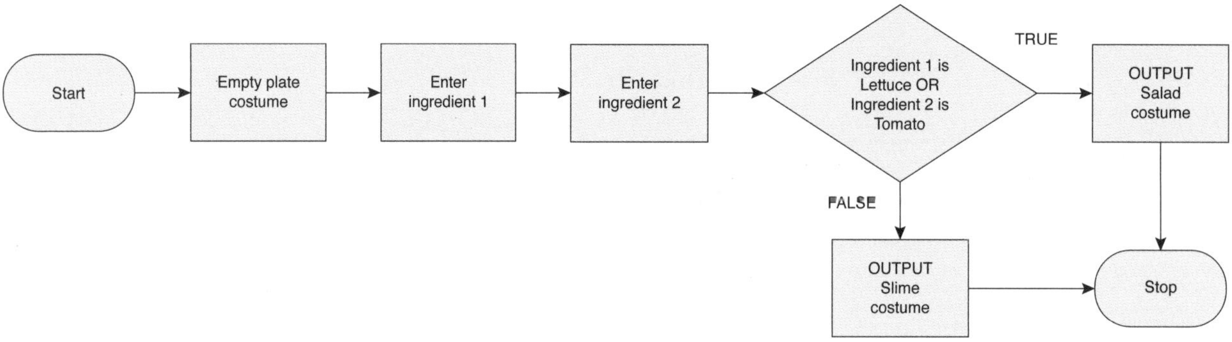

1 ....

....

2 ....

....

Chapter 3.4 Design a flowchart

Task A. Your feedback kiosk

1 Which location have you chosen for your feedback kiosk?

....

2 How will your sprite react to the rating?

- a costume change
- a costume effect
- an animation
- a sound
- something else (write below)

....

3 Design the algorithm for your feedback kiosk in the space below.

4 Turn your algorithm into a flowchart using the space below.

Return to page 46 of the Student's Book.

Reflection

Record the feedback that your partner gave you.

....

Homework

Find an example of a real feedback kiosk for a computer app or website. How does your example gather feedback?

....

Chapter 3.5 Build a program from a flowchart

Task A. Test your program

Check that your program works by completing the tests below. You should enter a number and record the output (reaction).

Number entered	Output
1	
2	
3	
4	
5	
6	
7	
8	
9	
10	

Return to page 47 of the Student's Book.

Reflection

Did your program work as intended? If not, did you fix any errors?

....

....

....

Homework

Show your project to a friend or family member. Ask them for feedback on your project and make notes on any improvements you would make if you had time.

....

....

....

....

Chapter 3.6 Showcase a flowchart and a program

Task A. Reflection

1 How accurate was your flowchart when you translated it into the Scratch code?

....

2 How accurate was your code when you completed the test plan?

....

3 What was your favourite part of your Scratch project?

....

4 Which part of your Scratch project would you improve if you had more time?

....

5 When you were viewing the projects that your class created, did you notice any features you would like in your own project?

....

Return to page 48 of the Student's Book.

Reflection

Share your favourite project made by someone in your class.

....

Homework

What changes did you want to make after showcasing your project?

....

....

What ideas from other projects would you like to add to your project?

....

....

Project skills checklist

- ☐ I can design a flowchart for a project.
- ☐ I can implement a flowchart in a Scratch program.
- ☐ I can make design decisions for a Scratch project.
- ☐ I can use logic statements in a Scratch program.
- ☐ I can test a Scratch project.

Chapter 4.1 Software and automation

Task A. System or application software

Place a tick in the table to state whether the software example is application or system software. The first one has been done for you.

Software name	System software?	Application software?
Word processor		✓
Antivirus		
Block-based programming software		
Online form creator		
Operating system		
Photo editing software		
Web browser software		
A computer game		

Key terms

Application software – Software used for a specific task, such as a word processor
System software – Software that controls how the computer operates
Operating system – Software that manages the hardware, software and resources of a computer
Utility software – Software that provides additional support for a system, such as antivirus or disk cleanup management
Automation – Carrying out tasks with little to no human interaction

Return to page 52 of the Student's Book.

Task B. Is automation good or bad?

Pages 53 and 54 of the Student's Book introduces automation in the health and advertising industries. Use this information to help you decide whether automation is good or bad. You will use your reasons in a class debate.

Write down two reasons why automation is good.

....

....

....

....

Write down two reasons why automation can have negative consequences.

....

....

....

....

Return to page 54 of the Student's Book.

Reflection

Which side did you agree with the most? Can automation only be good or bad? Could it be both?

....

....

....

Homework

Advertising in the world around you

Between now and your next computing lesson, look for adverts in the world around you. They could be on the TV, a streaming site, while browsing online, while playing a game, in a magazine or newspaper, on the radio, on bus stops or other public places.

Write down at least two adverts that you have noticed in the world around you. State whether you think automation was used to display the advert to you.

Advert	What was the advert and where did you see it?	Was it targeted to you automatically?
1		
2		
3 (Optional)		

Chapter 4.2 Binary representation

Task A. Count in binary

Your teacher has just demonstrated a Scratch project to you that shows how to represent the decimal numbers 0 to 10 in binary. Use what you learnt from this demonstration to write the decimal numbers 0 to 10 in binary. A table has been provided to help you with this.

Decimal	8	4	2	1
0				
1				
2				
3				
4				
5				
6				
7				
8				
9				
10				

Return to page 55 of the Student's Book.

Task B. Represent characters in binary

The table below shows all of the capital letters in the English alphabet and the binary codes that are used to represent each letter. Use the table to write the names of each of these in binary:

a your favourite animal

b your favourite colour

c your favourite activity.

A	01000001	**J**	01001010	**S**	01010011
B	01000010	**K**	01001011	**T**	01010100
C	01000011	**L**	01001100	**U**	01010101
D	01000100	**M**	01001101	**V**	01010110
E	01000101	**N**	01001110	**W**	01010111
F	01000110	**O**	01001111	**X**	01011000
G	01000111	**P**	01010000	**Y**	01011001
H	01001000	**Q**	01010001	**Z**	01011010
I	01001001	**R**	01010010		

a animal	
b colour	
c activity	

Task C. Represent sound in binary

An analogue sound has been recorded and needs to be converted into a digital sound. Use the image below to take samples of the wave. The first three have been completed for you. Note that you must choose the binary number that the sample is closest to.

> **Tip** You can mark the image with an X to help you to find the matching binary number.

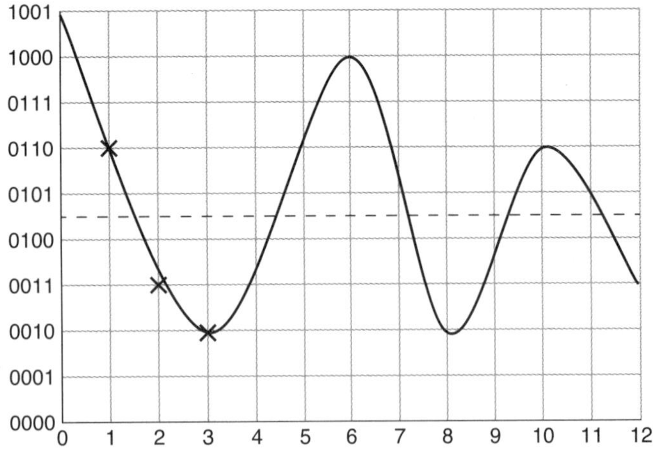

Time	Binary	Time	Binary	Time	Binary
1	0110	5		9	
2	0011	6		10	
3	0010	7		11	
4		8		12	

Return to page 57 of the Student's Book.

Reflection

How could a binary number be used to represent a single colour?

....

....

....

Homework

Your name in binary

Use the binary character table in Task B 'Represent characters in binary' to write your name in binary below:

Chapter 4.3 Represent images in binary

Task A. Draw the 1-bit colour depth images

The images below use 1 bit to represent the pixels in the image. This means that the images can only ever have two colours. 0 represents white and can be left blank. 1 represents the colour of your pen or pencil.

Follow the binary code in the table on the left to recreate the image in the table on the right.

0 – white, leave empty

1 – your pen or pencil colour

a

1	0	1	0	0
0	1	0	1	1
1	0	1	0	0
0	1	0	1	1
1	0	1	0	0

b

1	0	1	0	0
1	0	1	0	0
1	0	1	0	0
1	0	1	0	0
1	0	1	0	0

c

1	0	0	1	1
0	1	1	0	0
1	0	0	1	1
0	1	1	0	0
1	0	0	1	1

Return to page 58 of the Student's Book.

Task B. Colour combinations

If your image has a 1-bit colour depth, then it can have 2 unique colours.

If your image has a 2-bit colour depth, then it can have 4 unique colours.

Tip Look back in your book to the 'Count in binary' section. You can use this to help with this task.

1 Write down all the combinations of 1s and 0s that you can make with 3 bits.

....

....

....

2 How many colours in total can you use with a 3-bit colour depth image?

....

3 Write down all the combinations of 1s and 0s that you can make with 4 bits.

....

....

....

....

....

4 How many colours in total can you use with a 4-bit colour depth image?

....

Return to page 59 of the Student's Book.

Task C. Images that represent my community

Use the space below to sketch and write down ideas of images that represent your local community.

Return to page 60 of the Student's Book.

Reflection

What image do you think you might use in your project?

....

....

....

Homework

Spot the pattern

The table below shows how many combinations can be made with 1s and 0s of different bit depths. Can you spot a pattern?

Number of bits	1	2	3	4
Combinations	2	4	8	16

Answer:

....

Using the pattern, work out how many combinations you can make with 5 bits.

....

Chapter 4.4 Design your pixel art

Task A. Design three versions of your pixel art

For each design:

- Choose a colour palette.
- Draw your pixel art using the binary bit patterns.

Design 1

Colour palette:

000	
001	
010	
011	
100	
101	
110	
111	

Design:

Design 2

Colour palette:

000	
001	
010	
011	
100	
101	
110	
111	

Design:

Design 3

Colour palette:

000	
001	
010	
011	
100	
101	
110	
111	

Return to page 61 of the Student's Book.

Design:

Reflection

Do you have a favourite design from the three that you drew in this lesson? Were you happy with your colour palette? Did you have to change your colour palette as you were drawing the image?

....

....

....

Homework

Colour palettes in company logos

Every company has their own colour palette. For example, Google uses a palette that includes blue, red, yellow and green. You can see these colours in all of their applications.

Choose a company that you are familiar with and see if you can spot all of the colours that they use for their brand style.

1 Which company did you choose?

....

2 Which colours does the company use?

....

....

....

Chapter 4.5 Create your pixel art

Task A. Draw your final design

Draw your final design in the space below. Complete the colour palette for your design.

Colour palette:

000	
001	
010	
011	
100	
101	
110	
111	

Return to page 62 of the Student's Book.

Final design:

Reflection

How does your artwork represent your school community?

....

....

Homework

Favicons

When you visit a website, you might notice that a tiny little logo appears next to the name of the tag. This is called a 'Favicon'. Favicons are either 16 × 16 pixels, like the one you have been designing, or 32 × 32 pixels.

Explore three different websites and sketch the Favicon that appears, in the space below:

Website:	Website:	Website: ....

Chapter 4.6 Showcase your artwork

Task A. Reflection

1 How useful was the colour palette that you chose? Did you have to change it as you created your final design?

....

2 How well do you think your design represents an element of your local community?

....

3 What was it like to see your piece of art used as part of the full installation?

....

Return to page 63 of the Student's Book.

Reflection

What was it like to see your piece of artwork as part of a collaborative installation? How well did your class collaboration represent your community?

....

....

Homework

Research some famous pixel art images

Use your internet searching skills to find some examples of famous pixel art images.

There is a large community of creators who create digital art pieces using pixel art techniques. Pixel art is very common in retro game design.

Write down your findings below:

Project skills checklist

- ☐ I can choose a colour palette
- ☐ I can design a pixel art image
- ☐ I can create a pixel art image that is related to your local community
- ☐ I can finalise a pixel art image
- ☐ I can combine a personal pixel art image with those made by my class to make a large art installation

....

....

....

....

Chapter 5 Create with code 2

Chapter 5.1 Python: input and output

Task A. Print statements

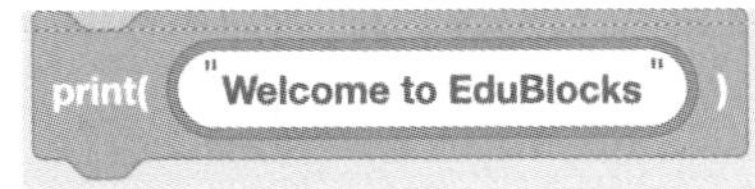

What is the correct Python code for this EduBlocks print statement?

In Python, code needs to be typed correctly. Click on the 'Code' tab in EduBlocks if you need help.

Code: ....

Return to page 68 of the Student's Book.

Task B. Silly sentences

Introduction

Your teacher just demonstrated a 'silly sentence' program that used the following code.

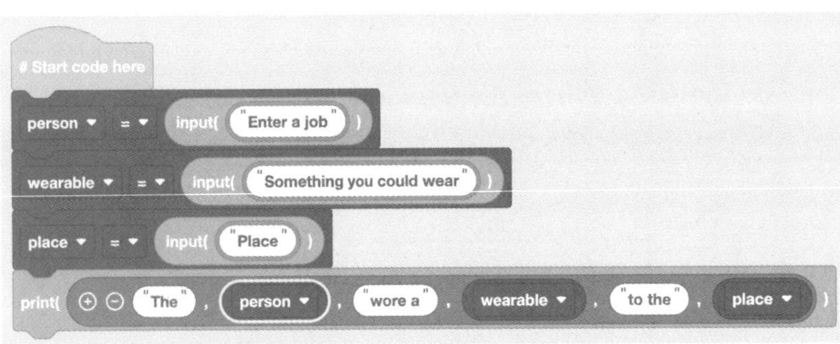

Your teacher will have run the code and entered three responses. The program will have displayed those responses in a silly sentence.

Below is some example output:

```
Enter a jobDriver
Something you could wearHat
PlaceKitchen
The Driver wore a Hat to the Kitchen
```

Your turn

Plan a sentence that has three blanks.

Example: The _____ wore ______ to the ______.

Your sentence:

....

....

What three prompts could you give your user to fill in the blanks?

Example prompts:

1 Enter a job

2 Enter something you could wear

3 Enter a place

Your prompts:

1 ....

2 ....

3 ....

What variables will you need? Your variables should have helpful names to match the data that they store.

Example variable names:

1 job

2 wearable

3 place

Your variable names:

1 ....

2 ....

3 ....

Build

Build your program using the example code given at the start of the task for support.

Here is an example output:

The <u>engineer</u> wore a <u>hard hat</u> to the <u>construction site</u>.

Write down an example output that your program could produce.

Your example (silly or sensible!):

....

Return to page 70 of the Student's Book.

Reflection

In what ways is Python similar to Scratch? In what ways is it different?

....

....

....

Homework

Order the code

A program is to be created that will ask a user to state their favourite colour and will then provide some text feedback. The blocks of Python code that the programmer will use are in the table below.

Show the correct order in which the code blocks should be placed by adding numbers to the 'Number order' column.

Code block	Number order
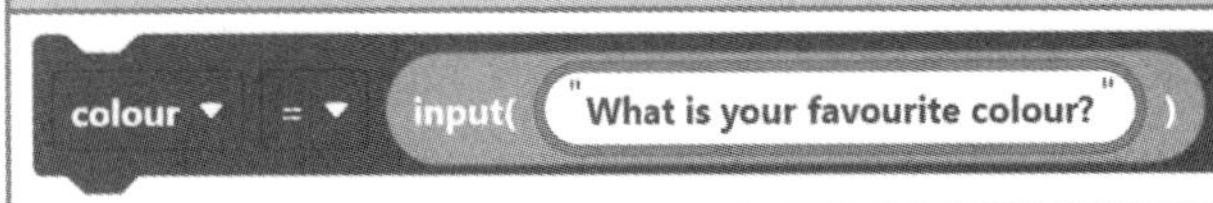	

If you have access to a computing device with the internet, then you could try the code blocks in EduBlocks to see if you were correct.

Chapter 5.2 Python: operators, conditions and data types

Task A. Why do we need data types?

> **Key terms**
>
> **Data type** – Classification such as string or integer that tells the computer how to work with a value
>
> **Integer** – Whole number

1 Enter the following code into EduBlocks by dragging blocks.

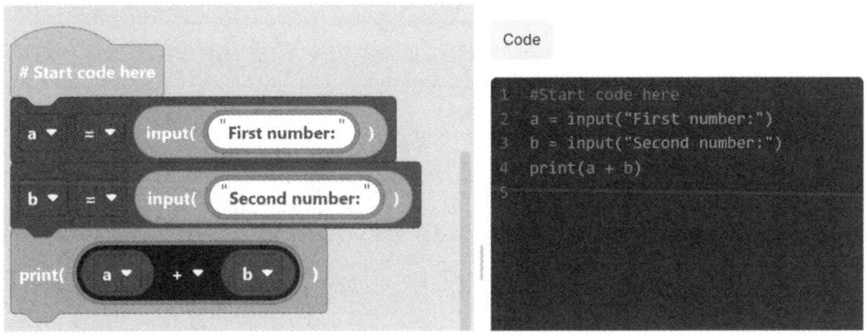

2 Predict what will be printed as output when you enter the numbers '6' and '7':

3 Run your code and enter the numbers '6' and '7'. What was the actual result?

4 Why do you think Python gave the result it did?

Return to page 71 of the Student's Book.

Task B. I need an int

Update your code from Task A to use the `int` function to convert the string inputs into numbers.

1 Use your code to calculate the answer to this calculation: 341267 + 932674

2 Change your code to calculate the answer to this calculation: 324 * 125

> **Tip** Python uses '*' to mean multiplication.

Return to page 72 of the Student's Book.

Task C. Using decimal numbers

Run the project from Task A and B and enter the numbers 4 and 2.5.

Write down what happens.

....

....

Return to page 72 of the Student's Book.

Task D. Calculate ticket prices

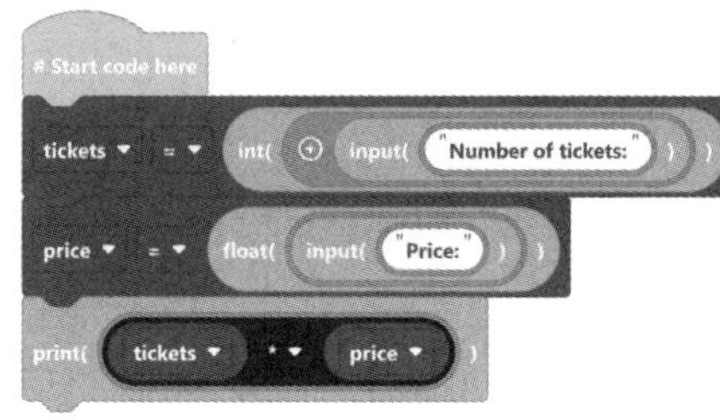

1 Practice using the 'float()' function by entering the code below into a new program in EduBlocks.

2 Use your new program to calculate the following costs:

   a 4 tickets at 4.5 each:

   ....

   b 12 tickets at 2.75 each: ....

   c 152 tickets at 0.75 each: ....

Return to page 73 of the Student's Book.

Task E. True or False?

Would these Python programs print True or False?

a
```
print(5 < 6)
```
....

b
```
print(12 != 10)
```
....

c
```
total = 10
money = 5
print(money >= total)
```
....

d
```
price = 2.5
amount = 4
total = amount * price
print(total == 10)
```
....

Tip You could check your answers using EduBlocks to see if you are correct.

Tip These questions have been written using the text version of Python. In EduBlocks you will also see the '# Start code here' block.

Return to page 74 of the Student's Book.

Reflection

Why are data types important in Python? What data types do you know and what values can they have?

....

....

....

Homework

Draw lines to match the Python blocks to their descriptions.

This block assigns a value to a variable.

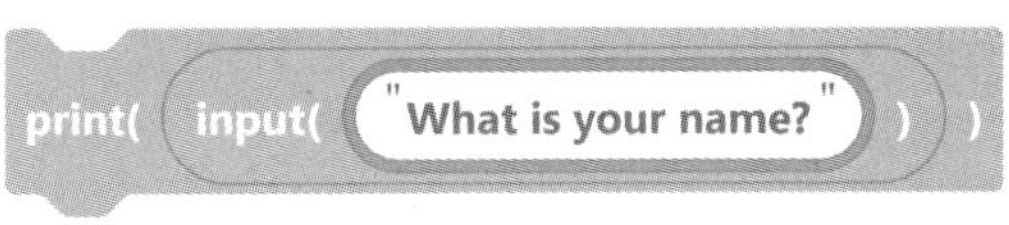

This block prints a message on the screen as output.

This block is used for selection, to check if a condition is True.

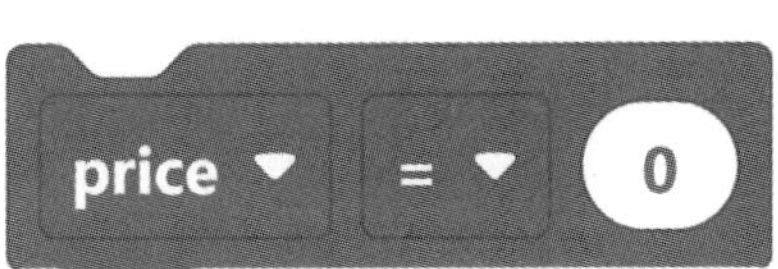

This block prints what is stored in a variable as output.

This block prints a message and accepts an input from the user.

Chapter 5.3 Python: Boolean logic, comments and functions

Task A. Decisions

Run the 'Book recommendation' project and enter different input values. Enter different values to get all of the possible book recommendations.

Fill in the values that you used to get the different recommendations.

Recommendation	Age	Likes
The Girl Who Stole an Elephant		
Maya and the Robot		
Zola's Elephant		
Blippy the Robot		

Return to page 75 of the Student's Book.

Task B. Comments

Code

```
1  #Start code here
2  age = int(input("How old are you? "))
3  likes = input("Do you prefer robots (r) or elephants (e)? ")
4  # Older child who likes elephants
5  if age >= 8 and likes == "e":
6    print("I recommend The Girl Who Stole an Elephant")
7  # Older child who likes robots
8  if age >= 8 and likes == "r":
9    print("I recommend Maya and the Robot")
10 # Younger child who likes elephants
11 if age < 8 and likes == "e":
12   print("I recommend Zola's Elephant")
13 # Younger child who likes robots
14 if age < 8 and likes == "r":
15   print("I recommend Blippy the Robot")
16
```

Notice that EduBlocks adds line numbers to Python code. These are not part of the program but are very helpful when we want to talk about a line of code.

Write down the numbers of the lines of code that are comments and will be completely ignored when the program runs.

....

Return to page 76 of the Student's Book.

Task C. Recognising patterns

Look at this code:

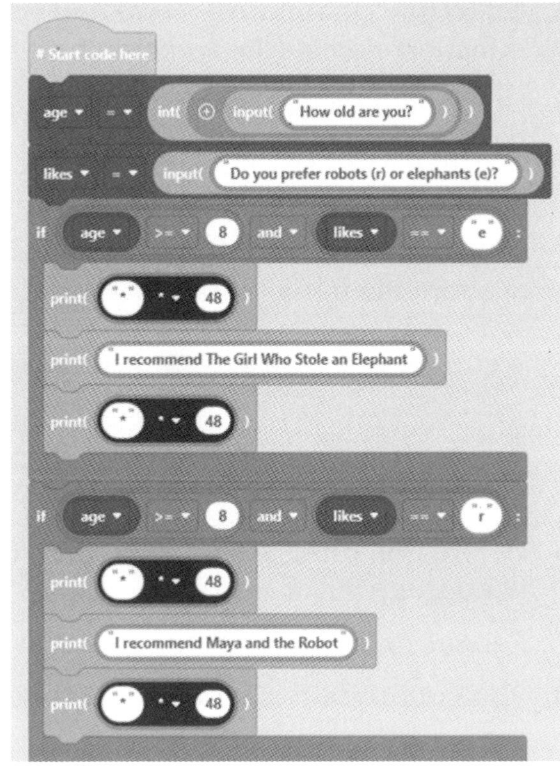

Consider the code *inside* the two if statements.

What is the same about this code and what is different?

Same: ....

....

....

....

Different: ....

....

....

....

Return to page 77 of the Student's Book.

Task D. Modify the recommendation app

Your teacher will give you a starter project that you will modify during this task.

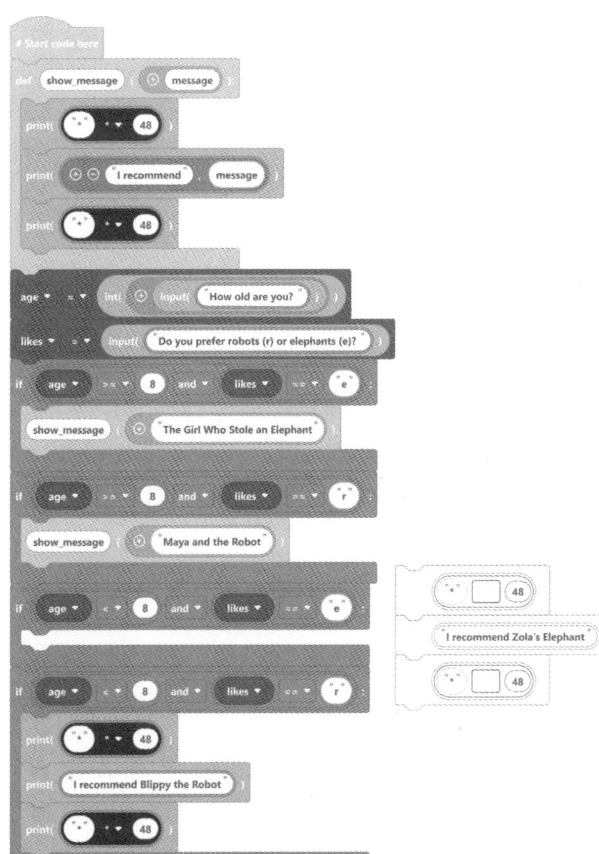

1. Clone the project to allow you to edit the code.
2. Find the third if statement in the program code.
3. Click on the first 'print' statement and drag all three lines of code to one side.

4 Go to the 'Definitions' menu and find the 'function_name' (the function call) block.

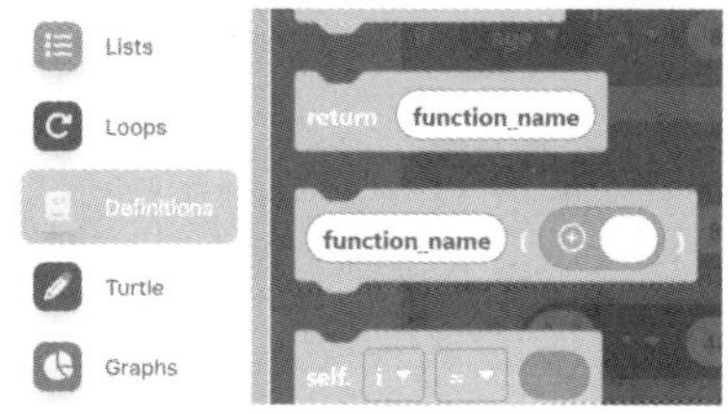

5 Drag the 'function_name' block inside the third if statement:

   a Change 'function_name' to 'show_message'.

   b Add the message "Zola's Elephant". This matches the book recommendation from the initial print blocks that you put to one side.

Tip Notice that "Zola's Elephant" has quotes around it. You will need to add these in yourself.

6 Delete the 'print' statements that you put to one side as you no longer need them.

7 Repeat steps 3 to 6 for the fourth if statement.

8 Run your code to check that it works as expected.

Return to page 78 of the Student's Book.

Task E. Debugging a function call

A learner has created a function call for the book recommendation app. They are testing their code and they get an unexpected error.

After running their code they typed in '9' for the age and 'e' to say that they liked elephants. The following text shows the output of this code.

```
How old are you? 9
Do you prefer robots (r) or elephants (e)? e
************************************************
I recommend I recommend The Girl Who Stole an Elephant
************************************************
```

The output message contains the words 'I recommend' twice.

1 Identify where the bug is in the code by circling the error.

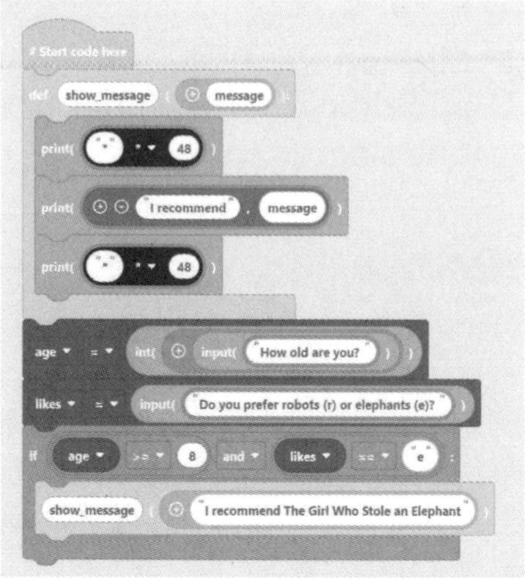

2 Write down how you would fix the error in the code.

....

....

Return to page 78 of the Student's Book.

Reflection

Did you have any errors in the 'Modify the recommendation app' task? If so, how did you fix them?

....

....

....

Homework

In your next lesson you will start to create your own recommendation app. To help you to prepare for this, think about the theme on which you would like to base your app.

It could be:

- a book recommendation app
- an activity recommendation app
- an after school club recommendation app.

Write your ideas below:

....

....

....

....

....

Chapter 5.4 Start your recommendation project

Task A. Plan your recommendation system

This is an example plan for your recommendation project.

Recommending: Books

Greeting: "Welcome to the book recommender," (name)

Number (float or integer): age

Text (string): what they like (robots or elephants)

Conditions	age < 8	age >= 8
Likes robots	Blippy the Robot	Maya and the Robot
Like elephants	Zola's Elephant	The Girl who Stole an Elephant

Variables needed: name, age, likes

Recommending:

Greeting:

Number (float or integer):

Text (string):

Conditions		

Variables needed:

....

....

....

....

Reminder

Your project should:

- greet the user using their name
- use at least one text input and at least one numeric input (integer or float)
- use conditional operators
- define a function and call it multiple times
- have helpful comments
- use helpful variable names.

Return to page 80 of the Student's Book.

Reflection

What errors have you found in your code and how did you fix them?

....

....

....

Homework

1 What problems did you encounter when you started to create your recommendation system?

....

....

2 Were you able to overcome the problems? If so, how did you do it?

....

....

Chapter 5.5 Complete your recommendation project

Task A. Project checklist

Tick off the tasks as you complete them:

- ☐ Create the greeting.
- ☐ Get the user input.
- ☐ Make decisions (complete the if statements) on which recommendation to make.
- ☐ Format the output nicely.
- ☐ Add comments.
- ☐ Test your code.
- ☐ Improve your code.

Return to page 81 of the Student's Book.

Task B. Meeting the brief

Tick off the objectives as your project meets them:

Your project should:

- ☐ greet the user using their name
- ☐ use at least one text input and at least one numeric input (integer or float)
- ☐ use conditional operators
- ☐ define a function and call it multiple times
- ☐ have helpful comments
- ☐ use helpful variable names.

Return to page 82 of the Student's Book.

Task C. Testing

Test your project. You should create tests that ensure that each recommendation is displayed when the conditions are met.

Fill in the values that you used to get the different recommendations. An example has been provided in the first row.

Recommendation	Number (input):	Text (input):
Example *"Zola's Elephant"*	*Example* *8*	*Example* *e*

Return to page 82 of the Student's Book.

Reflection

What are you looking forward to about sharing your project? Are you concerned about any parts of your project?

....

....

....

....

....

....

Homework

List three reasons why it is important to test your programs:

....

....

....

Chapter 5.6 Share your recommendation project

Task A. Feedback

1 Reviewer name: ....

2 List two things that you liked about the program:

....

....

3 List one thing that could be improved in the program:

....

4 Were there any errors with the code? If so, write them in the space below:

....

....

....

Return to page 83 of the Student's Book.

Task B. Reflection

1 Did you meet all of the project brief? Did it:

- ☐ greet the user using their name?
- ☐ use at least one text input and at least one numeric input (integer or float)?
- ☐ use conditional operators?
- ☐ define a function and call it multiple times?
- ☐ have helpful comments?
- ☐ use helpful variable names?

2 Describe one piece of feedback you got.

....

....

3 Describe how you used the feedback to improve your project.

....

....

4 Would you recommend your recommendation system? Why?

....

Return to page 83 of the Student's Book.

Reflection

How did you enjoy creating a recommendation system that could be used by others?

....

....

....

Project skills checklist

- ☐ I can plan a recommendation system
- ☐ I can modify code in EduBlocks/Python to meet the user's requirements
- ☐ I can test my code
- ☐ I can debug my code
- ☐ I can check that my project met the requirements
- ☐ I can improve a project based on feedback

Homework

Think about the recommendation program that you have created. If you had made it in Scratch, you could have used sprites, backdrops and costumes to provide recommendations to your user.

Draw a picture of what your recommendation program might look like if it was made in Scratch.

Chapter 6.1 Internet standards: DNS and HTTPS

Task A. Identify the parts of a URL

Your teacher will tell you which websites you can visit. Fill in the table below with example web pages that you visit. The scheme will be https if your browser says that the connection is secure.

scheme		domain name		web page
https	://	example.com	/	path/to/file.html

Return to page 87 of the Student's Book.

Task B. Top-level domains

The top-level domain for 'example.com' is '.com'.

What are the top-level domains for the examples you used in Task A?

1 ....

2 ....

What is the top-level domain for your country?

3 ....

Return to page 87 of the Student's Book.

Task C. The Domain Name System

In your own words, explain the purpose of the Domain Name System (DNS).

....

....

....

....

Return to page 89 of the Student's Book.

Reflection

Why is it important to have standards such as IP, DNS and HTTPS?

....

....

....

Homework

As you move around your community during the week, look out for top-level domains on signs, buildings, vehicles, adverts and so on.

Make a note of any that you see.

....

....

....

....

....

....

....

....

....

....

....

....

....

....

....

....

....

Chapter 6.2 Wireless data transmission

Task A. Which wireless technology?

Which wireless technology would you choose for each scenario and why? Refer to the feature comparison in the Student's Book.

Scenario	Wireless technology	Reasons for choice
Wireless headphones		
Internet of Things device for monitoring moisture in the garden or a home		
Alert system for elderly person that sends a message		
Robot programmable from an app		
A washing machine that can be controlled and monitored from an app		
A voice assistant device that can read audio books and control household devices using voice commands		

Return to page 90 of the Student's Book.

Task B. 5G features

Which 5G improvement would help with each of the following problems? Draw a line to connect each problem to its solution.

Jan wants to meet up with a friend during a busy celebration where lots of people are using their mobile phones in a small area.

Mobile phone usage has an impact on the environment due to power usage.

A sports fan wants their web page to load quickly so they can see the results.

A medical device needs to provide accurate and timely data in an emergency situation.

Faster data transfer speeds

Faster reliable communication for devices

Improved energy efficiency

Increased network capacity

Return to page 91 of the Student's Book.

Reflection

How do you use wireless technology in your life? Where have you seen wireless technology in your community? How have you been affected by problems with wireless technology? How would you make 6G better?

....

....

....

....

....

....

....

....

....

....

Return to page 92 of the Student's Book.

Homework

Move around with your wireless devices and observe how position affects the connectivity.

Make a note of your findings.

....

....

....

....

....

....

Student's Book page 93

Chapter 6.3 Encryption and data security

Task A. Decryption

1 The Caesar cipher encrypts messages by counting forwards or backwards by a number called the 'shift'.
Given the following plaintext and ciphertext messages, what is the shift?
Plaintext: THIS IS A SECRET MESSAGE

a Ciphertext: XLMW MW E WIGVIX QIWWEKI
What is the key? Shift:

b Ciphertext: RFGQ GQ Y QCAPCR KCQQYEC
What is the key? Shift:

2 Quickly remind yourself of the Pigpen grid in the Student's Book and then hide it. Redraw the grid here and then use it to decrypt this message.

Return to page 94 of the Student's Book.

Task B. Encryption

1 Which URL uses encryption? Tick the correct answer.

a http://example.com **b** https://example.com **c** secure://example.com

2 Write the missing word in each sentence.

a .... turns ciphertext into plaintext.

b .... turns plaintext into ciphertext.

Return to page 94 of the Student's Book.

Task C. Decrypt a message

Do not write the plaintext message here!

Swap books with a partner and write the ciphertext in their book for them to solve. Also write the algorithm and, if you chose the Caesar cipher, the shift number as the key.

Algorithm:

Key (if Caesar):

Ciphertext:

Plaintext:

Use the decryption algorithm and key to decrypt the message that your partner has written in your workbook.

Reflection

How much more difficult would it be to decrypt a message if you didn't know the encryption algorithm and didn't have the key?

....

....

....

Homework

Create a message and encrypt it using Atbash.

....

....

....

....

....

....

....

Explain the decryption algorithm for Atbash to someone else and ask them to use it to decipher your message.

Did you explain Atbash well enough so that they were able to decipher your message? How do you know?

....

....

Chapter 6.4 Create your own encryption algorithm

Task A. Plan your substitution cipher

What kind of letters or symbols will your substitution cipher use?

Make notes and sketch ideas.

Tip Look at the images in the Student's Book to remind yourself of some examples.

You will need 26 different symbols that you can create as Scratch sprite costumes.

Tip You can use the Text tool to enter text, including emoji characters. Ask your teacher how to do this on your school computer

A		B		C		D	
E		F		G		H	
I		J		K		L	
M		N		O		P	
Q		R		S		T	
U		V		W		X	
Y		Z					

Return to page 98 of the Student's Book.

Task B. Start creating your substitution cipher

Tick off each letter as you add a costume for it in Scratch.

A	B	C	D	E	F	G	H	I	J	K	L	M
N	O	P	Q	R	S	T	U	V	W	X	Y	Z

Write down any notes about your progress and plans that you will need for the next lesson.

....

....

Return to page 98 of the Student's Book.

Task C. Save and load your Scratch project

Give your Scratch project a sensible title and save it. ☐

Make sure that you can load the project. ☐

Make a note of where the project is saved and/or what it is called. (This could be the name of the Scratch account if you are working in a pair, or the location of the file on the school network.)

....

....

Return to page 98 of the Student's Book.

Reflection

Did you overcome any issues with your Scratch project? If so, share them.

....

....

Homework

Have a conversation about coded messages with people you know. Have they ever communicated using a coded message? With each person, discuss why they used a coded message, the rules and methods used and whether it was successful. Record your findings.

Who I asked: ....

Had they ever communicated using a coded message? ....

Comments: ....

....

....

Who I asked: ....

Had they ever communicated using a coded message? ....

Comments: ....

....

....

Who I asked: ....

Had they ever communicated using a coded message? ....

Comments: ....

....

....

Student's Book page 99

Chapter 6.5 Complete your encryption algorithm

Task A. Prepare a message

Think of a message to encrypt. It should be around 50 characters long (around 10 words).

Plaintext message:

....

....

....

....

Choose a team name to identify your code.

Team name: ....

Task B. Prepare a clue

Choose a word from your message to be used for a clue. The word should be at least 5–7 characters long.

Clue: Word number ....

is ....

Return to page 100 of the Student's Book.

Task C. Cracking your code

How easy do you think it will be to crack your code?

....

....

....

Do you think a clue will be needed? ....

Do you think you will need to provide the key? ....

Return to page 100 of the Student's Book.

Reflection

Have you ever cracked a code without knowing the encryption algorithm or key? How did you do it?

....

....

....

Homework

Create a piece of artwork that shows your name in your code.

Chapter 6.6 Cracking the code

Task A. Crack the code

Try to crack the code you have been given.

(If you are not given a code or you finish early then you can try to crack the code above.)

> **Tip** You can ask for a clue if you need it.

Use this space for notes to help you to crack the code.

....

....

....

....

....

....

....

....

....

....

Task B. Clue

Make a note of the clue if you need it.

Word number .... is ....

Return to page 101 of the Student's Book.

Task C. Plaintext message

When you have cracked the code, write the plaintext message here.

....

....

....

Return to page 101 of the Student's Book.

Task D. I cracked a code!

Describe the activity you completed in this lesson using the vocabulary you have learnt in this chapter.

....

....

....

....

Return to page 101 of the Student's Book.

Task E. Reflection

Now that you have created and cracked codes it is a good time to reflect on your process.

1 Give an example of a substitution cipher you used and explain the decryption algorithm.

....

....

....

2 Did anyone need clues to crack your code? If so, why did you give those particular clues?

....

....

....

3 How does knowledge of frequency analysis help to crack codes?

....

....

....

Return to page 101 of the Student's Book.

Reflection

Did you enjoy cracking the codes? Explain your answer.

....

....

....

Project skills checklist

- ☐ I can create my own encryption algorithm
- ☐ I can crack an encryption algorithm created by someone else
- ☐ I can relate my experience to modern internet security

Homework

Your teacher will give you a second printout of your code to take home.

Give it to someone else to solve.

Then complete the following:

Person I gave my code to: ....

Were they able to solve it? ....

How many clues did they need? ....

Student's Book page 104

Chapter 7.1 Modelling data to make decisions

Task A. Simulators in industry

Research the use of simulators in either the health, manufacturing or retail industry.

Key terms

Simulator – App or machine that realistically models real-world processes and conditions

1 Chosen industry: ....

Tip Use this area to make a note of keywords you think will help you with your research.

2 Describe how the simulator is used. ....

....

....

....

....

3 List your sources of information. ....

....

....

Return to page 105 of the Student's Book.

Task B. Data-driven decisions

Use your model to answer these questions.

1 Radio station 1

a What is the royalty amount per play? ....

b What is the average earned per day for the trial? ....

c How many days would it take, after the trial, to reach the target? ....

Return to page 107 of the Student's Book.

2 Radio station 2

a What is the royalty amount per play? ....

b What is the average earned per day for the trial? ....

c How many days would it take, after the trial, to reach the target? ....

3 Use the model to make a data-driven decision.

a Which radio station should the singer choose? ....

b What is a risk of this approach? ....

4 The singer agrees to go with radio station 1. They negotiate with the station and manage to get 0.40 in royalties per play. Update one cell in your model and then answer the following questions.

a Which cell did you update? ....

b How many days would it take to reach the target with the new royalty amount?

....

Return to page 108 of the Student's Book.

Reflection

Did you prefer seeing the data in the spreadsheet or represented graphically in the Scratch project? Why?

....

....

....

Homework

Think of an idea for a computer game that is based on a simulation. Describe the game and the data that it would use.

....

....

....

....

Chapter 7.2 Organising and formatting data

Task A. Identify primary keys

1 What issues are there with these chosen primary keys? The chosen primary keys have been circled in the tables.

a Date of birth ....

Student ID	Name	Date of birth	Subject
A1234	Tami	21.03.2012	Computing
A1235	Ola	28.02.2012	Science
A1236	Ike	21.03.2012	History

b Rank ....

Player ID	Name	Country	Rank
A1234	Ama	Ghana	1
A1235	Adjo	Ghana	
A1236	Chanda	Zambia	2

c Test score ....

Student ID	Name	Test score	Test percentage
A1234	Tami	54	60%
A1235	Ola	54	60%
A1236	Ike		

2 Circle the fields that would make the most appropriate primary keys.

a

Category	Subcategory	Product ID	Price
Books	Fiction	BK0001	10.50
Toys	Crafts	TY0001	10.50
Books	Non-Fiction	BK0002	6.50

b

Registration no.	Make	Year	Car colour
A01ABC	Hyundai	2020	Red
Z01XYZ	Suzuki	2020	Black
G50KLM	Hyundai	2021	Green

Return to page 111 of the Student's Book.

Task B. Highlight scores

A teacher has received a set of test results and wants to highlight all scores over 70%.

1 Create a new tab called 'Test results' and enter the data below.

Learner	Result
Learner 1	80%
Learner 2	90%
Learner 3	60%
Learner 4	70%
Learner 5	70%
Learner 6	80%
Learner 7	60%
Learner 8	70%
Learner 9	90%
Learner 10	90%
Learner 11	80%
Learner 12	80%
Learner 13	70%
Learner 14	60%
Learner 15	70%
Learner 16	60%
Learner 17	100%
Learner 18	70%
Learner 19	80%
Learner 20	60%

2 Create a single colour conditional formatting rule that highlights the entire row using the settings on the right.

3 How many of the class scored over 70%?

....

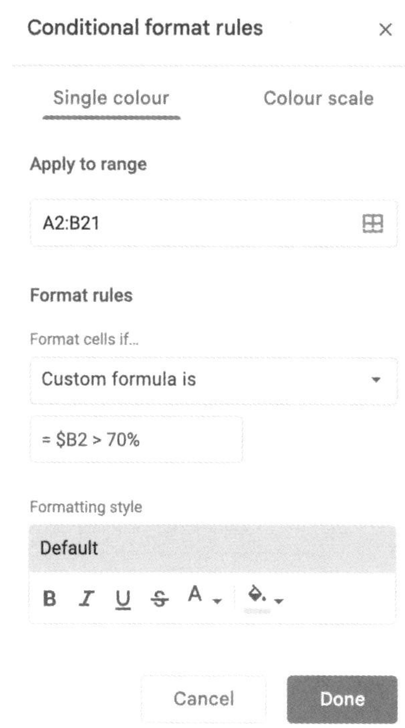

Return to page 114 of the Student's Book.

Reflection

Is highlighting the entire row useful or not? Why do you think this?

....

....

Homework

Create a spreadsheet of a collection of items you have at home. Include a primary key. Remember that a primary key must be unique and not empty. You can add a new field if none of the existing fields is suitable. For example, you could choose books, food items, toys, websites or music.

Write down the items you chose and the primary key. Include headings and an example row from your spreadsheet.

....

....

Chapter 7.3 Search and capture data

Task A. Searching databases

Name	Life expectancy	Diet	Wingspan
Parrot	30 years	Herbivore	19 inches
Penguin	20 years	Carnivore	16 inches
Albatross	50 years	Carnivore	120 inches
Hummingbird	5 years	Omnivore	4 inches

1 Which birds would be returned with the following database queries?

a Wingspan < 20 inches ....

b Diet = Carnivore ....

c Life expectancy > 20 years ....

d Wingspan = 16 inches ....

e Life expectancy < 5 years ....

Return to page 115 of the Student's Book.

Task B. Effective forms

1 Which is the best form? ....

2 Circle the features that make it the best form.

1

Invitation to Art Club

Progress

What is your name?
E.g. Divya Patel

Which days can you come?
Tick all the days you are available

- ☐ Monday
- ☐ Tuesday
- ☐ Wednesday
- ☐ Thursday
- ☐ Friday

Would you be interested in photography?

- ○ Yes
- ○ No — If no, skip next question

Do you have a camera?

- ○ Yes
- ○ No

2

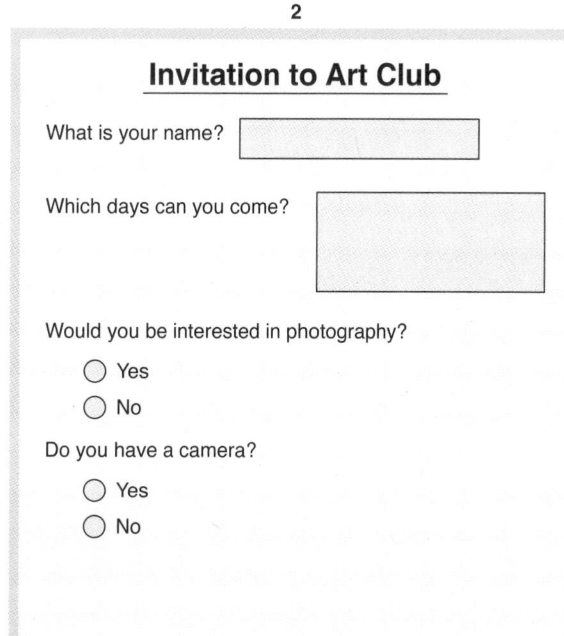

Invitation to Art Club

What is your name?

Which days can you come?

Would you be interested in photography?

- ○ Yes
- ○ No

Do you have a camera?

- ○ Yes
- ○ No

Return to page 116 of the Student's Book.

Reflection

Which tips did you use when creating your data capture form?

....

....

....

Homework

Ask a trusted adult about a time they had to fill in a data capture form. Did they find the form easy to use? Did they experience any problems?

Note: Make sure that you do not share any personal information about the adult in your answers.

....

....

....

....

Chapter 7.4 Ask your audience

Task A. Design your form

Tip

- Keep it short or use multiple page forms.
- Show progress as a percentage or visual graphic.
- Arrange questions in a logical order.
- Limit free-text fields.
- Use checkboxes and radio buttons. Use the 'Other' option if necessary.
- Use expected answer tips or descriptions.
- Use 'skip to section/question' functionality based on other answers.
- Add a confirmation message when the form has been submitted.
- Decide if the question is required or not.

Sketch and label key features of your form here.

Return to page 118 of the Student's Book.

Reflection

Did you have any challenges designing your form? If so, describe them.

....

....

....

Homework

Describe how you would improve your data capture form if you had more time.

....

....

Chapter 7.5 Analyse your data

Task A. Create your model

Tip Remind yourself about conditional formatting using page 111 of your Student's Book.

Use this area to make notes about the conditional formatting you will use in your model.

Return to page 119 of the Student's Book.

Task B. Student-led decisions

1 How many students did you survey? ....

2 What was your process for getting students to complete your form?

....

....

3 How will you spend the money?

....

....

....

4 How did you reach this decision?

....

....

....

5 What data did you add conditional formatting to?

....

6 How did conditional formatting help you to reach your decision?

....

....

7 How will conditional formatting help when you present your spreadsheet in the next lesson?

....

....

Return to page 119 of the Student's Book.

Reflection

Share the challenges that you had while creating your model.

....

....

....

Homework

Create a presentation with 1–3 slides to support the showcase in the next lesson. Explain how you used best practice to design your data capture form and how you used conditional formatting to help to analyse your data.

Chapter 7.6 Present your decision

Task A. Reflection

Now that you have presented your findings and seen other presentations, it is a good time to reflect on your process.

1 Did you ask the right questions in your form?

....

2 Which questions would you like to improve or add if you were to do the project again?

....

3 Why was your conditional formatting scheme effective for:

a making data-driven decisions

....

b presenting the reason for your decisions?

....

4 What would you like to improve or add to your spreadsheet?

....

Return to page 120 of the Student's Book.

Reflection

Why are spreadsheets useful tools for making data-driven decisions?

....

....

Project skills checklist

- ☐ I can create a spreadsheet model for a royalties calculator
- ☐ I can make decisions about data based on modelling
- ☐ I can design a capture form
- ☐ I can use a capture form

Homework

Summarise the experience you have gained during this chapter. How will it make you think differently next time you fill in or design a data capture form?

....

....

....

Chapter 8.1 Project plans

Task A. Order the tasks

A new app is being developed to allow students in a school to access their homework schedule and track their progress. It is currently January, and the project needs to be completed by August. It should be fully tested by real students in a school before launch. The app should allow students to:

- be notified when there is a new piece of homework
- read details of the homework
- submit homework
- see the grade that they achieved for their homework.

Key terms

Project plan – Carefully planned tasks that ensure a project meets a deadline to the correct requirements

The homework app has not yet been started, but the project manager has written a list of the main tasks that will need to be completed to meet the deadline. Read the list of tasks and decide on the best order in which to complete them.

Tip Indicate the order by adding numbers in the final column. The first one has been completed for you.

Task	Order
Consider the design options for the app.	
Build the app.	
Test the app with students.	
Meet with all stakeholders to decide on the scope of the project.	
Test the app internally.	
Establish a budget for the project.	
Launch the app.	
Create a timeline for the project.	1

Return to page 124 of the Student's Book.

Task B. Case study: Mobile ticketing

Managers of an event venue had a meeting about sustainability. They agreed that to meet UN Sustainable Development Goal 12 they would massively reduce their use of plastic.

They decided that the only way to access the venue for concerts and sports events would be via the use of mobile tickets on smartphones. They created a new app where users could buy tickets and download them straight to their smartphones.

The first event was a sports match. Fans who had attended matches for years could no longer attend because they didn't have smartphones.

To make things worse, the project cost double the allocated budget as new features of the app and upgraded ticket reading machines had been added during development.

There were two main reasons why this project was unsuccessful. What were they?

....

....

Return to page 125 of the Student's Book.

Task C. Effective design

An internet company is creating a new search engine. During interviews with customers, they found out that they:

- access the search engine from their mobile phones
- use only one hand to hold their phone and search.

Based on this information, they have created three possible designs.

1 Circle the design that is best suited to the needs of these customers.

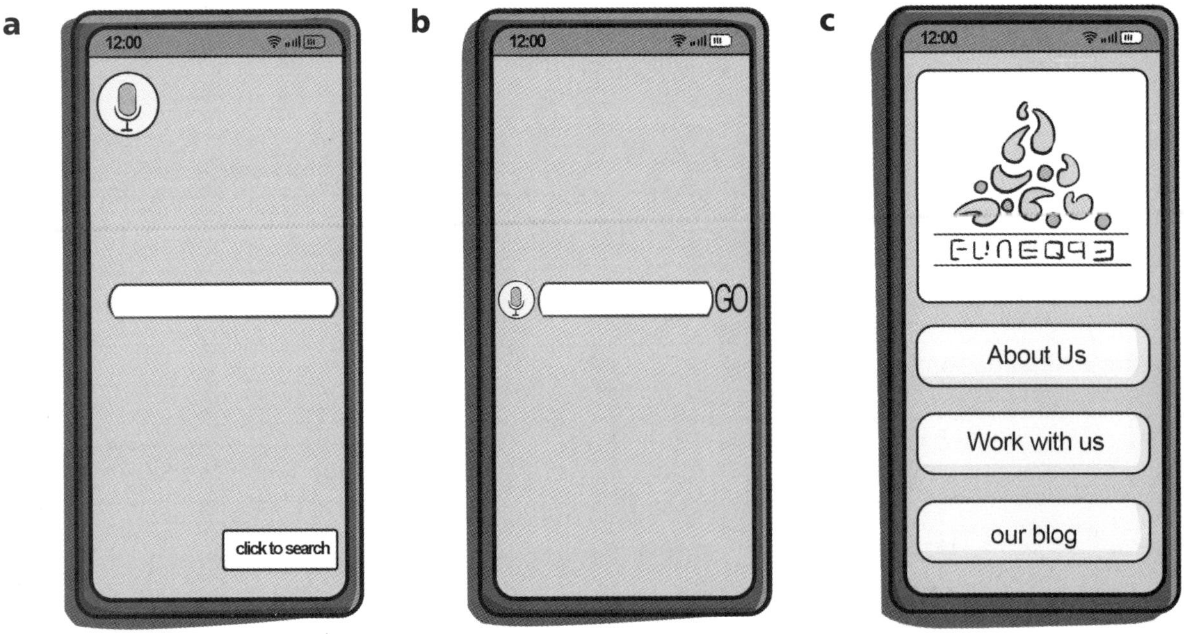

2 Why is the design you circled the best option?

....

....

Return to page 126 of the Student's Book.

Reflection

Share the key things you learnt from this lesson. What did you find most interesting?

....

....

....

Homework

Interview a friend or family member about an app or device that they use.

Write down two things they like about the app or device.

1 ....

2 ....

Write down two things they think could be improved.

1 ....

2 ....

Do you think the app or device developer did a good job of understanding user requirements? Why or why not?

....

....

....

....

Chapter 8.2 Test plans

Task A. Follow a test plan

Your teacher will provide you with the EduBlocks program that you need to test.

Key terms

Test plan – List of tests to be carried out on a finished program to ensure that it works as expected

A student developed a micro:bit addition device. The device:

- displays a smiley face as a welcome message
- generates a random number between 10 and 100 when a user presses button A
- generates a random number between 10 and 100 when a user presses button B
- adds the two numbers together and then scrolls the answer.

The student also created a test plan.

1 Run the code in the micro:bit simulator and document the results and fixes. You will need to:

A Click on the 'Simulator' button.

B Press the 'play' icon on the micro:bit.

C Use the buttons on the micro:bit labelled 'A' and 'B'.

D From Accelerometer drop-down menu, select 'shake' then click 'send'.

#	Description	Input	Expected output	Result	Fix (if needed)
1	A welcome message displays	Switch on / play	A smiley face icon		
2	User generates the first number	A	A random number between 10 and 100 is shown		
3	User generates the second number	B	A random number between 10 and 100 is shown		
4	The result displays	Shake	The result of adding the two numbers together is shown		

Return to page 128 of the Student's Book.

Task B. Write and follow a test plan

A student created a micro:bit multiplication device. The device:

- displays "Hi" as a welcome message
- generates a random number between 1 and 12 when a user presses button A
- generates a random number between 1 and 12 when a user presses button B
- multiplies the two numbers together and then scrolls the answer.

1 Create a test plan.

2 Run the code in the micro:bit simulator to document the results and fixes.

#	Description	Input	Expected output	Result	Fix (if needed)
1					
2					
3					
4					

Return to page 128 of the Student's Book.

Reflection

Why is testing important?

....

....

....

Homework

Research a famous software error from a space mission. Describe what happened in your own words and how testing could have avoided the problem.

....

....

....

....

....

....

Chapter 8.3 Prototypes

Task A. Does it meet the requirements?

The toy manufacturing department of a global company has been asked to develop a virtual pet for children. It should:

Key terms

Prototype – Sketch or model to test a concept for a project

- be able to be switched off
- allow a user to choose which animal they will have as a pet from three different options
- when switched on:
  - ask for food every 2 hours
  - ask to play a game every 3 hours.

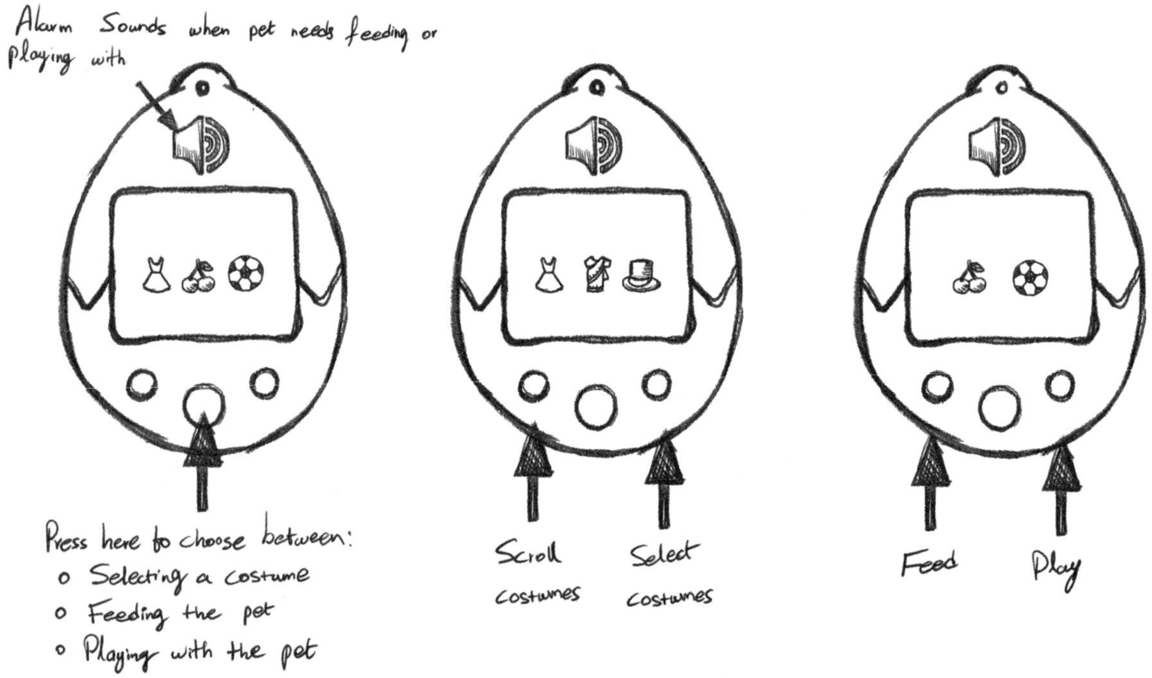

Look at the prototype sketches that have been provided.

1 Have the requirements been met?

....

2 If not, what is missing and how could the design be improved?

....

....

....

Return to page 129 of the Student's Book.

Task B. Your prototype sketch

1 Create a prototype sketch for your 'classroom welcome' device in the space below. Remember to label your features.

Ask for feedback about your sketch. Does it meet the requirements? Could any more features be added? Will you be able to build your project in the time that you have? Note your feedback below:

....

....

Task C. Your improved prototype sketch

Using the feedback that you have just received, redraw your prototype and provide new annotations.

Return to page 130 of the Student's Book.

Reflection

Why is prototyping important when creating a new product?

....

....

....

Homework

Choose one of the following scenarios and describe how the change in requirements would affect the design for your device.

a The device needs to be suitable for visually impaired visitors to your classroom.

b The device is to be used to welcome visitors to a library.

....

....

....

....

....

....

Chapter 8.4 Create your device

Task A. Plan your build

Use this page to write notes about your code. You could:

- write the names of the image or music settings you want to use
- map your inputs to your outputs
- draw a grid to sketch any custom display image you want to make
- note which libraries you will need to import.

Return to page 131 of the Student's Book.

Reflection

How successful were you at sticking to your prototype without scope creep?

....

....

....

Homework

A friend suggests that it would be brilliant if your device could respond to five different inputs and have three animations and play tunes that you write yourself.

Would it be a good idea to add all of these additional features at this point in the project? Why or why not?

....

....

....

....

Chapter 8.5 Test and refine your device

Task A. My test plan

Create your test plan in the table below, use your success criteria to help you. Use your test plan to test your device, document the results and note the fixes that you need to make.

#	Description	Input	Expected output	Result	Fix (if needed)
1					
2					
3					
4					
5					
6					
7					
8					
9					
10					

Return to page 132 of the Student's Book.

Task B. User research

Test your device with at least two end users.

Observe their interaction with your device:

- When they use the input does it trigger the output?
- Do they have positive reactions to the outputs?
- Does their body language tell you anything?

Ask the users for feedback:

- Did they understand what they had to do?
- Do they think that your inputs were easy to use?
- Would they feel welcomed when using your device?

My observations	User feedback

Return to page 132 of the Student's Book.

Reflection

What differences did you notice between testing your device in the simulator and testing your physical device with end users?

....

....

....

Homework

Write instructions for using your device. Make your instructions clear for a user who has never seen a micro:bit before.

....

....

....

....

....

....

Chapter 8.6 Showcase your device

Task A. Reflection

1 Did your project meet all of the requirements?

....

2 Did your test plan include everything you needed to properly test all aspects of your code? Did you find any problems after testing? What tests should you have included to avoid this?

....

3 Over this entire Stage, you have performed many showcases. Do you think that you have improved at giving a showcase across the year? If so, in what way?

....

4 If you had more time, what would you change about your final design?

....

5 Which 'classroom welcome' device from your class did you like the best and why? You can't choose your own!

....

Return to page 133 of the Student's Book.

Reflection

How well did your device create a welcoming environment for visitors to your classroom?

....

Homework

Find an app or device that you use regularly. What feature do you think the product development team should add next? Write a letter to the developers requesting the new feature and explaining why they should add it. Remember that they will be working to a budget and schedule and will want to avoid scope creep.

You do not need to send the letter. Talk to a trusted adult first if you do want to send the letter.

....

Project skills checklist

- ☐ I can plan a project
- ☐ I can prototype ideas
- ☐ I can build a physical computing device
- ☐ I can test a device against a test plan